THE TEACHING OF PHYSICS IN SECONDARY SCHOOLS

MAXWELL PRESS

THE TEACHING OF PHYSICS IN SECONDARY SCHOOLS

John F. Woodhull

Submitted in Partial Fulfillment of the Requirements for the Degree f Doctor of Philosophy in the Faculty of Pure Science, Columbia University

MAXWELL PRESS

Chennai   Trichy   New Delhi

MAXWELL PRESS

This edition has been published in india by arrangement with Carson Books, UK

ISBN 978-93-91270-02-5   **Maxwell Press**

All rights reserved   No. 44, Nallathambi Street, Triplicane, Chennai 600 005

MJP 1075   © Publishers, 2021

Publisher :   C. Janarthanan

Publisher's Note

The legacy of a country is in its varied cultural heritage, historical literature, developments in the field of economy and science. The top nations in the world are competing in the field of science, economy and literature. This vast legacy has to be conserved and documented so that it can be bestowed to the future generation. The knowledge of this legacy is slowly getting perished in the present generation due to lack of documentation.

Keeping this in mind, the concern with retrospective acquiring of rare books has been accented recently by the burgeoning reprint industry. Maxwell Press is gratified to retrieve the rare collections with a view to bring back those books that were landmarks in their time.

In this effort, a series of rare books would be republished under the banner, "Maxwell Press". The books in the reprint series have been carefully selected for their contemporary usefulness as well as their historical importance within the intellectual. We reconstruct the book with slight enhancements made for better presentation, without affecting the contents of the original edition.

Most of the works selected for republishing covers a huge range of subjects, from history to anthropology. We believe this reprint edition will be a service to the numerous researchers and practitioners active in this fascinating field. We allow readers to experience the wonder of peering into a scholarly work of the highest order and seminal significance.

Maxwell Press

JOHN FRANCIS WOODHULL, A. B.

Yale College, 1880.

Teacher in Secondary Schools, 1881–'85.

Student in Chemistry, Harvard Summer School, 1884–'85.

Student in Chemistry and Physics, Johns Hopkins University, 1885–'86.

Teacher of Science, New York State Normal School, 1886–'88.

Lecturer in National Summer School, 1888–91.

Professor of Physical Science, Teachers' College, Columbia University, 1888– .

Lecturer in Martha's Vineyard Summer School, 1890.

Student in Physics, Harvard Summer School, 1892.

Lecturer at Chautauqua Summer School, 1894.

Student in Physics and Chemistry, Columbia University, 1898–'99.

ii

PUBLICATIONS.

1. Home-made Apparatus.   Popular Science Monthly, 1889.
2. Simple Experiments for the School Room.   E. L. Kellogg & Co., 1889.
3. Academic Syllabus for Physics and Chemistry.   Regents' Bulletin, No. 5, 1891.
4. Selection and Use of Apparatus.   Regents' Bulletin, No. 6, Part III, 1891.
5. Object Lessons, with David Salmon.   Longmans, Green & Co., 1892.
6. First Course in Science.   Henry Holt & Co., 1893.
7. Educational Value of Natural Science.   Educational Review, 1895.
8. Manual of Home-made Apparatus.   E. L. Kellogg & Co., 1895.
9. Systematic Work in Nature Study.   Regents' Bulletin, No. 36, 1896.
10. The Proper Use of Laboratory, Library, and Lecture in Teaching Physical Science in Secondary Schools.   Regents' Bulletin, No. 42, 1897.
11. Physics.   New York Teachers' Monograph, 1898.
12. Chemical Experiments, a Laboratory Manual, with M. B. Van Arsdale. Henry Holt & Co., 1899.
13. Physical Experiments, a Laboratory Manual, with M. B. Van Arsdale. D. Appleton and Company, 1899.
14. Physics.   A Text-Book, with C. Hanford Henderson.   D. Appleton and Company, 1899.

CONTENTS.

I. INTRODUCTION.

Since the Committee of Ten of the National Educational Association issued its report in 1892 upon the studies in secondary schools, it has been very much the fashion to discuss courses of study. It would appear, however, that while teachers of other subjects have organized and discussed and, in a great measure, agreed and prepared definite plans and carried them into execution, the teacher of science has been contented to do his work as an individual, heroically, it may be, but alone. Hence there is no settled practice as to the sequence of the various subjects in science and little or no attempt at relating them either to one another or to other subjects in the curriculum—no uniformity in opinions as to what a course in physics should really constitute, or what its aim should be.

It has seemed to me that here was a proper field for research, and that a dissertation might well be prepared on the teaching of physics in secondary schools in which an attempt should be made to determine the proper place of the subjects in the course, what should be its chief aim, and what its contents.

II. THE PLACE OF PHYSICS IN THE HIGH-SCHOOL CURRICULUM.

In discussing this question I shall give more weight to the needs of the pupil than the individual preference of teachers or the logical arrangement of the subject. Let us, however, ask whether these considerations are antagonistic to one another.

In logical sequence, I believe the physical sciences come

first, and biology and geography follow. This order is re-quired both for the sake of the subject-matter and for the sake of the method.

First, as to method. The study of all science requires the use of the laboratory method. I am satisfied that dur-ing the first few months of high-school work in science special attention should be given to training pupils in good laboratory habits, and pupils in the early part of the high-school course are at the right age—about thirteen to fif-teen years—to begin to learn the mechanical side of lab-oratory work. For this purpose I prefer chemistry in the first year, followed by physics in the second year. It is possible to put nearly all of the chemical apparatus in the hands of each pupil. Test tubes, beakers, and flasks con-stitute most of this equipment, and need not cost more than two or three dollars per pupil. This is not true of physics. No one expects to have physical apparatus enough to supply each member of his class with every piece, and hence the pupil in physical laboratory work must be a little more capable of self-direction on that account. The mechanical part of laboratory work in chemistry may, and I believe it should be, made simpler than that of physics. When the experiments are going badly in the laboratory, as they frequently will with beginners, the teacher should perform an experiment in the presence of the class, discussing the difficulties as he proceeds, and when he has finished have them all try the same experi-ment. In this manner he should train the pupils to do things as they ought to be done and to see things as they are. I should first teach them to bend a delivery tube that approaches pretty closely to some model taken as a standard. I should train them to transfer powders to test tubes without getting any upon the tables, or even upon the sides of the tubes; to invert a bottle of water for catch-ing gases without admitting any air; to heat test tubes and flasks without breaking them; to use acids without spotting the tables or the clothing or the labels upon the bottles; to generate chlorine, hydrogen sulphide, etc., without contaminating the air of the room. If they are not so trained when they are quite young they are hopeless. An old student can not be trained to become a good surgeon; he will let the bacteria in somewhere. The college professor who gets hold of a lot of careless students finds that he is quite helpless. He can do very little toward reforming their laboratory habits. Indeed, while the first high-school class will submit very willingly to this training in painstaking, careful habits, the third and fourth year pupils are already getting beyond that stage.

Next, as to subject matter. I put chemistry before physics, not only because it enables me better to introduce young people to laboratory methods so far as the mechanics of these go, but also because it may, and I believe it should, be treated as a simpler subject than physics with reference to training in thought power. We use Remsen's Elementary Course in Chemistry and Carhart and Chute's text-book in Physics. Remsen's book is not so difficult as Carhart and Chute's book. It is true that portions of both these books are not understood by the pupils. Most of the portions of the chemistry, however, which the first year high-school pupils do not comprehend will not be comprehended by the fourth year high-school pupils either, and these portions may very properly be left for the college course. Chemical theory has very little place in the high school in whatever year the subject is taught. We may treat the subjects so that *combustion* in chemistry shall be simpler than *heat* in physics; the *chemistry of water* simpler than the *mechanics of fluids*; the chemical conduct of chlorine, sulphur, nitrogen, carbon, etc., simpler subjects of study than electricity and sound. I find that chemistry lends itself very readily to the training of young pupils of thirteen or fourteen years of age, not only in the mechanics of the laboratory, but also in the more important process of making experiments lead to orderly thought. Pupils of that age can be readily taught all the chemistry necessary to fit for college, so that we need not be hampered by that consideration. This has been done in the Horace Mann School for the past ten years. The work is certainly no more difficult than the algebra which they are taking the same year. I regard chemistry as the key to all the sciences in the high-school curriculum.

Physical science, including chemistry and physics, furnishes the proper foundation for the study of biology and geography in respect both to method and to subject matter. Neither biology nor geography furnishes such good opportunities for training in the mechanics of the laboratory, nor do they furnish simple inductions. We may select our facts in physical science and arrange them so as to make the conclusions to be drawn from them as simple as we choose. Nature performs many physical experiments in a large way; these, as encountered in physical geography, may be too large for the youthful mind to encompass. We may perform in the physical laboratory experiments which represent the large phenomenon piecemeal, and thus build up the conception of the whole, and at the same time develop the constructive imagination which is needed to grasp the processes of nature. Often we may devise for the laboratory a miniature representation of the larger phenomenon of nature which shall be quite a perfect analogy. This not only shows the possibilities of laboratory work in chemistry and physics, but I believe it indicates in a general way a very important aim to be kept in view; viz., to select as far as possible such work as should prepare the way for the biology and geography to follow. I should teach chemistry and physics in the high school, not solely, perhaps not chiefly, for their own sake, but rather for the sake of furnishing a foundation for other work. Professor Tarr says: "Geography is a complex of all the sciences." So also is physiology, we may say. Now physiology runs through all biology. Whatever may be said of the study of plants and animals in the elementary school, in the high school I believe it should be chiefly the study of animal and plant physiology; and physiology deals to a very considerable extent with the chemistry and physics of living things. Mr. James E. Peabody, in the preface of his excellent laboratory manual of physiology, says: "Since physiology unfortunately precedes physics and chemistry in the ordinary high-school courses of study, it is necessary to give the pupils some idea of the fundamental principles of these subjects." Foster and Shore, in the preface of their little book called Physiology for Beginners, say: "A sound knowledge of physiology cannot be gained without some acquaintance with chemistry and physics, and at least a rudimentary knowledge of these ought to be obtained before the study of physiology is even attempted."

The report of the Committee of Ten states it in these words: "The study of physiology is in a great measure the study of the mechanics, the physics, and the chemistry of the living body; before it can be pursued profitably the student should have, at least, a fair elementary knowledge of these sciences as fundamental. It is not possible to teach it as science to pupils devoid of such knowledge."

Mr. Franklin W. Barrows, teacher of zoölogy in the Central High School, Buffalo, N. Y., states the case for zoölogy as follows:

"The time for the course in zoölogy should be, if possible, not earlier than the third year in the high school, and the pupil should have completed his laboratory course in physics and chemistry. It seems to me that any teacher who has had practical experience with laboratory work in other sciences as well as in biology, would prefer that his students should lay the foundation of their science work in the study of physics and chemistry rather than in biology. Many teachers of zoölogy and botany are attempting less with their classes than they would like to attempt, because they are constantly reminded of the limitations of their pupils—the untrained eye, the unskilled hand, and that disregard for the exact truth which incapacitates so many young people for any original or accurate accomplishment in science. Then again the problems of biology, involving the consideration of animal function and structure, require a greater degree of intellectual maturity than those of physics and chemistry. For this reason there is an immense advantage in postponing this work to the third or fourth year instead of plunging into it at the beginning of the course."

So much for logical sequence. Now, is there anything in the nature of the pupil which demands a different order of subjects than this? If so, I yield to that demand as imperative; but I have not been able, by nineteen years of experience in teaching the various sciences to high-school classes, and by careful consideration of that which has been written and spoken upon the matter, to discover any reason why this is not also the order of subjects which is demanded by the best interests of the child. It has been said that green things and living things appeal most to the interests of the pupils, and this is given as a reason for putting their study first in the course. I have been at a good deal of pains to examine into this matter, and I do not find it so. On the other hand, my evidence, gained from the children themselves, is overwhelmingly in favor of the statement that all pupils are very much more curious about what they may see in a chemical or physical laboratory than in all out doors. But there is a very wide difference between being interested in things and being interested in the formal study of things. On this score, however, if it is necessary or desirable to keep pupils at a white heat of interest, the teacher of chemistry or physics has the advantage. This bidding for the interest of the pupil is, nevertheless, in my judgment, a very undesirable proceeding. I assume that any capable teacher can do his work as it should be done without the necessity of competing with some one else for the interest of the pupils. If pupils *are* more interested in one subject than another, that ought not to be the only argument, nor the principal argument, for putting that subject in a certain place in the curriculum. A wholesome interest in a subject on the part of the pupil I regard as essential to his success, but a good teacher of any one of the sciences need have no lack of interest. After all, the argument of interest is all in favor of the logical sequence of the subjects. One great aim of science teaching is to accustom the mind to logical arrangement; to make one thought grow out of another; to so arrange things that one phenomenon shall appear in its relation to another; so that one fact may throw light upon another; to organize our everyday observations. The logical sequence not only of topics but of subjects is just what is needed to gain and hold the interest of the pupils.

It has been said that it is not the aim of teachers of botany and zoölogy in the first year high-school class to teach science, but rather to teach observation. This is called a faculty, and it is alleged that it will be useful to the student when he takes up science later in his course, possibly in college. This is compared to whetting one's tools for future service. One might as well say that it is not the aim of history teachers to teach history in the early part of the high-school course, but rather to teach memory.

Many books of object lessons have been published for the purpose of teaching observation to children as a mechanical process, so that they might have the faculty all ready to use when they get older.

This is stupid and stultifying business. A person think-ing much on some question, the answer for which he ex-pects to find in nature, is liable to see things which cast light upon his problem, while at the same time he may be very unobserving of those things which do not help toward the answer of his question. This is scientific ob-servation. Children are natural observers. They only need to be taught to select for a definite purpose, and to connect one observed fact with another. This, I imagine, is what Professor Huxley meant when he said, " Science is merely organized common sense." The number of observations which a person may make in a given time is often inversely proportional to the amount of thinking he does. A child and his father go to walk. By the roadside they both notice some object as they pass along. The father, having had a wider experience, finds something in this object which starts a train of thought. The child, not being so occupied, sees numerous other objects which the father fails alto-gether to notice or, if he does notice them, they flit across his mind without making any impression; but presently an object attracts his attention, and because it is related to the first it attracts his attention more than it does that of the child. The number of objects seen by the child during that walk may be to the number seen by the father as ten to three, but the father has seen related things; he has seen to some purpose; he, of the two, is the scientific observer; yet, if some one asks whether they passed a certain gate, the boy will be very likely to be the only one who can answer. Now, no teaching can enable the boy to see more things, but training may enable him to think more about what he sees. Some one has said that a good memory re-quires a judicious forgetfulness; so scientific observation requires that the mind shall not be distracted by observ-ing a multitude of unrelated things. Hence the teaching of scientific observation brings us back again to the logical sequence of subjects. This logical sequence in a high-school course I hold to be: first year, chemistry; second year, physics; third year, biology; fourth year, geography.

High-school pupils may properly be expected to work, in school and out, five and a half hours a day. Or, taking the school period of forty minutes as the unit, eight periods a day. Five of these periods may be taken in school and three may be devoted to study at home.

The pupils in the Horace Mann High School attend school thirty-two weeks in the year, and the course is four years in length. Eight periods a day, five days in the week, thirty-two weeks in the year, for four years make a total of 5,120 periods of work in the whole course. I believe one fifth (one quarter is recommended by the Committee of Ten) of this time, or 1,024 periods, should be devoted to science, and I recommend the following scheme:

	Periods per week in school.	Periods per week out of school.	Whole number periods per week.	For the year of thirty-two weeks.
First year, chemistry...........	4	2	6	192
Second year, physics...........	7	3	10	320
Third year, biology............	7	3	10	320
Fourth year, geography	4	2	6	192
				1,024

POSTSCRIPT.—It being desirable to consider whether the above arrangement would leave the course well balanced, I append herewith a schedule showing a fair allotment of time to other subjects.

	"WEEK HOURS."		Total in school and out.	Total No. of periods in school and out for four years of thirty-two weeks each.
	Prepared.	Unprepared.		
English and history	27........	0 =	40........	1,280
Science.......................	10........	12 =	32.......	1,024
Mathematics	11........	0 =	22........	704
Manual training and drawing....	0........	20 =	20.........	640
Modern languages..............	13........	0 =	26........	832
Latin	10........	0 =	20........	640
				5,120

III. WHAT SHOULD CONSTITUTE A YEAR'S COURSE IN PHYSICS FOR PUPILS FOURTEEN OR FIFTEEN YEARS OF AGE.

First, we ought not to attempt to teach any part of the subject completely. While I should regard the high-school work in physics as an almost essential preparation for the general course in physics in college, it should not be expected to complete any portion of the subject so as to make college work in that portion unnecessary. So far from the high-school work taking the interest out of the college work, I should expect it to increase it fourfold, and, so far from its creating in the student bad habits either of doing or of thinking which will require to be corrected in college, I should expect it to beget habits essential to good work in college; and if the pupils fail to receive this training in the high school, I should expect it would be very difficult for the college instructor to accomplish wholly satisfactory results with them. I should leave out entirely much that is usually found in a high-school course, making, however, the portion which is left a complete and logical whole. I should expand, much more fully than it is the custom to do, those topics which appeal most to the pupils. Such topics should be wrought out to the fullest extent in their application to physical geography, meteorology, astronomy, biology, physiology, cooking, the phenomena of the kitchen, the factory, the shop, the farm, etc. This does not interfere with the more extended study of physical geography, physiology, cooking, etc., afterward. It gives the best possible introduction to them, and at the same time makes the physics mean something. Physics is too often like a course in grammar. Too much attention is given to learning the definitions of things not immediately useful. A syllabus of topics for laboratory work in a second year high-school class is presented herewith (pp. 15–16). This must, of course, be amplified by lectures, text-book and recitation work. Some portions of physics which are important to give to high-school pupils do not lend themselves well to the laboratory treatment. Such portions will be found in the syllabus of topics for lecture and recitation work which I present on pages 39–40. As a rule laboratory work should go hand in hand with lectures and recitation work, otherwise neither is brought to fruition. Laboratory work is too often merely mechanical. Students may become adept at working with instruments of precision without gaining any outlook upon physics, without acquiring any habits of induction, or the faculty of seeing relations. I take it that the training of the constructive imagination is equally important with training in accuracy. The proper relation of laboratory, text-book work, and lecture I have stated in Regents' Bulletin No. 42, and quote it here:

" *The Laboratory.*—The chief purpose of the laboratory, in the high schools at least, is not for the exercise of the inductive method. Fascinating and fruitful as that method is in the hands of a certain few ripe scholars, it is often a snare and a delusion in the hands of the young or of the untrained adult. With great care pupils should be trained in the process of induction for its own sake, but not with the idea that that is the method by which they are to be taught any considerable portion of science. There is small chance for induction in the teaching of high-school chemistry or physics. Mechanics and light furnish the best opportunity for the use of that method. The chief purpose of the school laboratory should be to make knowledge real. Merely to see a piece of apparatus aids immensely one's appreciation of the physical principle which the apparatus illustrates, but when one manipulates the apparatus himself he acquires a knowledge of physical science which he can in no other way attain. In order that a laboratory exercise may be productive of good, two things must be looked out for: first, the pupil's mind must be carefully prepared for it beforehand. Nothing is more common than to find pupils going through their laboratory exercise in a mechanical way without acquiring much knowledge of the physical principles involved; and second, the pupil's mind must be guided to some definite end. This in most cases can be done by quizzes following more or less closely the laboratory exercises; but frequently, to prevent desultory and aimless work, it is necessary to conduct quizzes either of individuals or of the whole class during the laboratory exercises. High-school pupils can not with profit be left much to themselves in laboratory work. They need very specific written or printed directions. They must also have a discreet teacher, who will keep very busy supervising the work of each pupil while in the laboratory.

" *The Text-book.*—The earlier text-books were faulty in that they covered too wide a range of subjects and dealt too much with paradoxes. They seemed in many cases to be a deliberate attempt to make things seem unnatural and unreal. There was the hydrostatic paradox and the culinary paradox, the Cartesian diver or bottle imp, the " fountain in vacuo," the " Hero's fountain," the " Bacchus illustration," the " bolt head," etc. These were the topics around which the instruction centered, and these formed the items of examination questions. The text-books to-day have discarded much of this material, but they have become extremely meager and uninteresting. They still deal with too many principles, but treat them too meagerly. A principle is stated in a brief paragraph. This paragraph is followed by a description of a single experiment, and then another principle is served up. In many cases that which occupies half a page might better be expanded to cover a dozen pages. Silvanus Thompson's Elementary

Lessons in Electricity and Magnetism illustrates fairly well a good form of expansion of the text. But in addition to such matter as is to be found in this book, and perhaps with the elimination of some of it, I should recommend filling out the text-book with what might be called adipose tissue—the abundant illustration of each principle with well-told stories of incidents in everyday life—something calculated to make the pupil see physics in the world around him. Such expansion of the text-book would make it necessary to have it bound in several volumes—say one on heat, one on electricity and magnetism, one on sound, and one on light, etc. The pupil has an armful of books for Greek or history or English; why not for physical science? It must be so if the latter is to take its place among the former as a disciplinary and a culture study. As in history or English so in physical science, there must be extensive use of the library. Topics should be assigned which require the student to consult books of reference, not merely for tables of boiling points, etc., but the reading of magazine articles, of encyclopedias, and of volumes like Faraday's Chemical History of a Candle and Tyndall's Forms of Water and Heat a Mode of Motion.

"There should be beside text-books and books of reference a laboratory manual containing very specific directions for the work to be done in the laboratory, a great deal of information concerning apparatus and experiments, a great many queries, mathematical problems, and supplementary exercises, etc. This book should be well supplied with blank pages upon which the pupil may disclose the state of his own mind.

"*The Lecture.*—The strongest feature of good instruction in physical science is the lecture. The teacher of physical science should aspire to some reputation as an interesting lecturer. He has the best possible subject with which to attract an audience; why should he neglect the opportunity to interest and instruct and inspire by this method? Let him not despise the giving of university extension lectures for the good it may do himself. If he has learned how to enjoy the study of physical science in the laboratory, in the library, and in nature, let him inspire his students by giving them through his illustrated lectures a glimpse of his own broader view upon the subject. By lectures he can make physics and chemistry real, live subjects as nothing else can possibly do."

There are certain chapters in physics which must be treated chiefly in lecture and recitation; the laboratory work plays only a small part. There are two means of adjusting in such cases: first, we use some of the laboratory period for extra lectures and recitations; and second, we introduce some laboratory work which, although needful, may be irrelevant. This is the case in the early part of the course, where the laboratory work is upon the metric system and mensuration, while the lectures and recitations are dealing with properties of matter, gravitation, motion, etc. During a portion of the study of heat and sound the time of lectures and recitations is increased at the expense of laboratory work. During a portion of the study of the dynamics of fluids and electricity the laboratory work is increased at the expense of the lectures and recitations.

Laboratory work is so apt to be a blind following of directions without much thought upon the relation of the work, that we find it necessary to sprinkle among the written directions a great many queries and problems intended to lead the pupil out from the observation upon the individual experiment in hand to kindred observation in other experiments or in nature, that he may acquire the habit of looking for relationships and of making broader generalizations.

Laboratory work should require as good solid thought as does a recitation; indeed, there must be considerable of the recitation feature about it. To accomplish this the division must not number more than twenty-five, and the instructor must have an assistant capable of helping in his work of going about among the pupils while at their work, and giving the right kind of guidance and of asking the right kind of questions. This assistant should also be assigned the large and important task of examining critically the pupils' laboratory note-books, and discussing with each pupil such matters as need correction. Such assistant might very well be a recent graduate from the course in which instruction is to be given.

Syllabus of Topics for Laboratory Work in Physics for Second Year High-school Class.

I. Matter, Force, Mechanics of Solids.

Extension and Weight of Matter.—Metric System.

Determine the ratio between an inch and a centimeter by measuring distances varying in length from four inches to four feet. Use a meter stick and measure to half a millimeter; also a stick having inches divided to tenths, and measure to the quarter of a tenth.

In similar manner determine the ratio between an ounce and a gram, between a quart and a liter. Weigh a measured quantity of water and calculate the weight of a cubic centimeter of it in grams.

Determine the ratio between the circumference of a circle and its diameter or radius. Determine the ratio between the area of a circle and the square of its radius by cutting these figures in cardboard and weighing them. Find the volume of a given cylinder. Find the volume of a block of wood whose faces are perpendicular to one another. Find the surface of a sphere and its volume.

Experiments to illustrate cohesion, capillarity, crystallization, diffusion, and osmose.

Experiments with levers, from which problems may be drawn, which shall apply to pulleys, wheel and axle, cog-wheels, whiffletrees, joints of the body, steelyards, center of gravity, shears, nutcrackers, etc.

Uniformly Accelerated Motion.—Determine whether a ball rolling down an inclined plane will go three, five, and seven times as far during the second, third, and fourth seconds (or periods of time) as during the first second (or period of time).

The Second Law of Motion.—Determine whether an object projected horizontally will reach the floor in the same time as one let fall from the same height.

The Path of Projectiles.—Plot the path of a ball thrown upward in such a manner that it will have a vertical velocity of two hundred and fifty-six feet per second, and at the same time a horizontal velocity of forty-eight feet per second.

Pendulum.—Determine whether the time of vibration is independent of the length of the arc and the weight of ball but dependent upon the length of the pendulum. Determine the ratio of length to the time of swing. Determine the height of the laboratory ceiling by the time of vibration of a pendulum suspended from it.

II. Mechanics of Fluids.

Make a U-shaped bend in a glass tube, having one arm of the U about four inches and the other arm about fourteen inches long. Put a few inches of mercury into this tube, and fill the long arm nearly full with water. With a meter stick determine the relative weight of water and mercury.

Using mercury in a U-tube as a pressure gauge, determine whether the pressure in water is proportional to the depth without regard to the size or shape of the vessel which contains it; also whether at any given depth it is the same in all directions.

Make a scale for the mercury pressure gauge, translating the height of mercury column into pounds per square inch of pressure.

By means of the mercury pressure gauge determine the pressure at the water faucet in pounds per square inch. From this calculate the height of the water in the tank above the faucet. Will this calculation hold good if water is running from some other faucet? Put your mouth to the pressure gauge and find the pressure in pounds per square inch which you may exert with your lungs. Find the pressure on the steam pipes if there is a pet valve at the radiator. With water in a U-tube find the pressure of the gas in grams per square centimeter in the gas pipe.

Put various liquids, such as water, alcohol, or ether, in a flask, and close it with a rubber stopper having two holes. Through one hole put a thermometer, and through the other hole put a glass tube and connect by means of a rubber tube with the mercury pressure gauge. Warm very gently and calculate the tension in pounds per square inch which corresponds to certain temperatures.

Find the area of the cross-section of a glass specimen tube in square centimeters. Put shot into this so that it will float in a vertical position in a tumbler of water. Measure in centimeters the depth to which it sinks in the water. Calculate what is the buoyant force upon the bottom of the tube. What must be its weight? What volume and weight of water does it displace? Find its weight by using the balance. Find the volume of its submerged portion by floating it in a graduated cylinder.

A floating body displaces its own weight of the liquid. A body that sinks displaces its own volume of the liquid. The buoyant force equals the weight of the floating body. The buoyant force equals the weight of the liquid displaced. The buoyant force at any given depth is the upward pressure at that depth.

Weigh a marble in air. Weigh it when suspended in water. How much does the water buoy up? What is the volume of the marble? What is its weight compared with that of an equal volume of water (specific gravity)?

Find the volume of the marble by immersing it in a test tube containing water and computing the volume of the water displaced.

Balance a tumbler of water upon a scale pan. Weigh the marble, then suspend it in the water so as not to touch the bottom or sides of the tumbler, but have it completely submerged. Find how much weight must be added to the other scale pan to restore the balance. How much did the water buoy up? How much did the thread sustain? What is the volume of the marble? What is its specific gravity?

Illustrate in various ways that the air has weight, and that it exerts an equal pressure in all directions.

By means of a long mercury pressure gauge, the long arm of which is closed and free from all air, determine the pressure of the air in pounds per square inch. Read the height of the mercury on several successive days, and compare with official reports for the same hour and the same vicinity (barometer).

Verify Boyle's law for the relation of volume and pressure in the air.

By means of the mercury pressure gauge prove that a pressure exerted upon a fluid inclosed in a vessel is transmitted equally in all directions, the total pressure upon the walls of the vessel being proportional to the area.

Show how the principle is illustrated in the exercises which follow.

Using a single pressure gauge, find how much pressure in grams per square centimeter you can exert with your lungs; connect two pressure gauges " in parallel," and note the pressure as indicated by each; connect three pressure gauges " in parallel," and note the pressure as indicated by each.

Make the same tests, connecting the pressure gauges " in series."

Fasten a tube air-tight in the side of a jelly-cake tin, or shallow tin pan. Tie rubber cloth over the top. Lay a board upon this and place several pounds of weight upon it. Force air through the tube from your lungs. Remember how many pounds per square inch of pressure you were able to exert with your lungs in a former exercise (page 17), and account for the large weight which you are able to lift in this experiment.

Operate various pieces of interesting apparatus, such as the syphon, the Cartesian diver, etc., and discuss in laboratory note-books the applications of the principles of fluid pressures to them. Give definite problems with reference to each, such as the following:

In a large wide-mouthed bottle of water a small vial is inverted, having just enough air in its upper part to barely float it. The weight of the vial is 3.5 grams, and the air which it contains is 2.5 cubic centimeters. What must be the volume of glass in it, and what is its specific gravity? Tie rubber cloth air-tight over the mouth of the large bottle. If now we exert a pressure of five pounds per square inch upon this rubber what will the volume of air in the vial become? Will it sink? If so, with how much force will it press upon the bottom of the large bottle? If now we remove the pressure from the rubber cloth and pull upward upon it with a force of five pounds per square inch what will the volume of air in the vial become? Will the vial rise above the surface of the water in the large bottle? If so, what volume will emerge?

III. Heat.

I. HOW HEAT IS PRODUCED.

A. *Heat Produced by Friction.*

Grind a piece of iron upon a grindstone until it is hot enough to light a bit of phosphorus upon a plate. Hot axle on cars, on wagons; machinery warms up while running. What is done to obviate this? Saw gets hot while sawing wood. Grease used to reduce friction and thereby diminish heat. The charring of a stick of wood by rubbing against a rapidly moving wheel.

Produce sparks by striking a pebble a glancing blow against a rasp.

Sparks produced by emery wheel, horse's shoe striking the pavement, scissors grinder, car brakes.

Try to light a friction match by drawing it across, first, a very smooth surface (e. g., a pane of window glass), and second, a rather rough surface (e. g., a pane of ground glass). Why does it light easier upon one surface than the other? Rocks fused by flexing of the earth's crust.

B. *Heat Produced by Pressure.*

With a pump condense air in a bottle containing a thermometer to indicate changes of temperature.

With forceps pinch the rubber delivery tube to check the flow of steam from a flask in which water is boiling slowly. Have a thermometer in one hole of the stopper. Do not force the temperature above 105° C.

Notice difference in temperature in the steam which comes from two flasks, one of which is open while the other has a delivery tube nearly closed. Try the temperature, both with the hand and with the thermometer. Steam jet from locomotive. Fall of temperature when a " soda-water " bottle is uncorked.　Manufacture of ice.　Liquid air.　Internal heat of the earth.　Heat of the sun.

C. *Heat Produced by Chemical Action.*

Hold in the palm of one hand a small dish containing a piece of quicklime as large as a hen's egg.　Add cold water and note the heat which will be produced.　Give plenty of time and note how the heat increases.　Use the thermometer and see if the temperature rises as high as the boiling point of water ($100°$ C.).

Slow decomposition of organic matter; pile of green grass; grain elevator; cotton baled with the seeds left in; piles of oily cotton waste; hay put into a barn before it is properly cured; animal heat.

II. SOME EFFECTS OF HEAT.

Heat acts like a force within a mass to drive its particles farther apart.　It expands gases, liquids, and solids.

Heat a flask filled with air and closed with a rubber cloth.　Take a tumbler from hot soap solution and invert it upon a plate.

Hold the bulb of the air thermometer in the palm of your hand for a short time and note the height of the liquid in the tube.　Rub the palms of your hands together vigorously, and again close one of them around the thermometer bulb.　Note the height of the liquid in the tube. Hot-air engines.

Heat changes solids to liquids and liquids to gases by expansion.　Cups of paraffin and wax.

Heat a little water to boiling in a stoppered test tube. Steam engines.

Find what is the temperature of your blood upon either Fahrenheit or Centigrade scale, and translate it into the corresponding temperature upon the other scale.　Do the same with the temperature of the air in the room, the water from the faucet, or other specimens of water which may be furnished.

Note the size of bore of the tube of a broken thermometer. What is the advantage of having the tube so small and the bulb so large? What advantage in thin walls for the bulb?

Contraction by cold is in the following case more than balanced by expansion of crystallization.

Fill a one-ounce narrow-mouthed bottle with water, cork it, and surround it by a mixture of ice and salt in a tumbler. In fifteen or twenty minutes the water should have frozen and broken the small bottle. Why does ice float upon water?

The expansion of crystallization of various substances—type metal—such metals as may be cast. Why paraffin and wax candles may not be cast. See cups of paraffin and wax which have been melted and cooled.

Illustrate the expansion of solids by heating a wire. Examine a metal thermometer. Setting wagon tires. Cracking lamp chimneys. Allowance made for expansion by heat in the case of iron bridges, etc. Unequal expansion of different metals.

Apparatus for regulating temperature of room automatically.

Evaporation.—Warm a plate and put it by the side of a cold plate. Put a drop of water upon each and note the time of evaporation in both cases.

Heat aids evaporation. Dishes taken from hot rinsing water dry quickly; not so if the water is cold. Clothes dry more rapidly when rung out of hot water than cold water; when hung near a hot stove than in a cold room.

Make a solution of salt in water (about half a teaspoonful in an ounce of water) and distill a portion of it. Taste the solution and the distillate. Did the distillate pass through an invisible state in the process of distillation?

The process of distillation is illustrated by what goes on in the laundry. The water evaporates from the boiler and is condensed to drops upon the window pane. This process goes on in a large way in nature. The evaporation from ocean, and lakes which have no outlets, must be equal to all the water that is supplied to them by rivers and rain. Fog, mist, clouds, rain, and dew.

Dew Point.—In a bright metal cup put water and add slowly little pieces of ice. Stir with a thermometer and notice the temperature of the water at the moment dew begins to appear upon the outside of the cup. Slowly add water, stir continually, and notice the temperature of the water at the moment the dew disappears.

Drops of water upon an ice pitcher. " Seeing the breath " in a cold atmosphere.

Arrange the following according to the facility with which they evaporate at the temperature of the room: ether, alcohol, water, gasoline, benzine.

See if you can cause mercury to evaporate by heating it in a test tube.

Some substances may pass directly from the solid to the gaseous state without apparently passing through the liquid state.

Heat iodine in a test tube. The evaporation of camphor. The evaporation of ice on cold, *windy* days. Clothes " freezing dry."

Relation of pressure to evaporation.

Breathe into two eight-ounce flasks so as to cloud them with moisture. Pump air from flask and compare the evidences of evaporation in the two flasks.

Temperature of water in engine boilers; extraction of gelatin.

Put a little water in a flask and close it with a rubber stopper having two holes. Put a thermometer through one hole and a delivery tube through the other. Heat the water nearly to the boiling point, remove the lamp, and connect the delivery tube with the exhaust nipple of a pump. Take a few strokes and note the effect upon the rate of evaporation. Also note the temperature as indicated by the thermometer. Exhaust still further, and see if you can make water boil at a temperature, say, of 70° C.

Boiling point on mountain tops. Vacuum pans in sugar refineries, milk condensers, etc.

Effects of currents of air upon evaporation.

Evaporate drops of water from two plates, blowing across one of them meanwhile.

Clothes dry more quickly upon windy than upon still days. "Muggy days." Humidity.

III. HOW HEAT IS DIFFUSED.

A. *Heat Diffused by Conduction.—Conductors, Good and Bad.*

Heat at their conjunction an iron and a copper wire which have been linked together. Pass the hand along each wire toward the flame to determine which is the better conductor of heat.

Burn a match at one end, no heat is perceptible at the other. Melt a short piece of glass at one end while holding the other end in the hand.

Fasten a piece of sheet metal to a board and cover the whole with a sheet of paper. Heat in the flame the paper at the point of the junction of wood and metal. How does sheet zinc protect the woodwork underneath and behind the stove?

Heat two iron wires in the flame—one a very thin wire and the other a thick one. Why does the small wire quickly reach the melting point of iron, while the larger one is scarcely heated red hot?

Hold in the flame thin wire gauze made of the same wire as above. Why is it not easily heated red hot as the single wire was? Why does not the flame pass up through the gauze? By means of a lighted match determine whether unburned gas is passing up through the gauze.

Extinguish the flame of the bunsen burner, lay the wire gauze over the mouth of the lamp, light the gas above and slowly raise the gauze. Why does not the flame extend downward to the mouth of the lamp? Miners' safety lamp.

Float an evaporating dish upon a vessel containing a small amount of water. Pour alcohol into the evaporating dish and light it. By means of a thermometer immersed in the water, determine whether the water conducts heat. Stir the water and note the reading of the thermometer now. Does the summer's sun heat bodies of water to any great depth?

Put a small thermometer into a test tube and heat the upper end of the tube in the flame. Does the air in the tube conduct heat to the thermometer? Remove the flame and tip the tube so that its mouth shall be slightly below the other end. Note the change in the thermometer.

Conducting and nonconducting materials, used for clothing; used for handles of stove utensils; wrappings for hot stones, for furnaces, for steam pipes. How are the crops protected by snow in winter? Double windows used in houses in winter time.

B. *Heat Diffused by Convection.*

Cause a current of air to pass through a pasteboard box by means of two argand lamp chimneys and a lighted candle.

Place a small flame under one side of a beaker with a metal partition in it. Drop in flakes of something to show convection currents in water.

Illustrate how currents of water move to and from the hot-water tank in the household.

Apparatus for heating buildings by hot water. Winds. Ocean currents. Ventilation.

C. *Heat Diffused by Radiation.*

How do we get heat from an open fireplace? Would we be able to receive heat from this fire if there were no atmosphere or matter of any kind in which conduction or convection could occur? Transparency for heat.

How do we get heat from the sun?

IV. SPECIFIC HEAT AND CALORIMETRY.

Have two hundred cubic centimeters of water at the temperature of the room in each of two vessels (calorimeters). • Note the temperature. In a third vessel have two hundred cubic centimeters of water with a piece of lead in it weighing two hundred grams. Have a thread attached to the lead so that you may transfer it from one vessel to another. Cover this third vessel to prevent loss of water by evaporation, and heat it nearly to the boiling point of water. Remove the lamp, take the temperature of the water (and lead?) and quickly transfer the lead to one of the aforementioned vessels of water and into the other pour the hot water. Stir the hot and cold water together, and move the lead about for a few moments to make the temperature in each uniform. Now note the resulting temperature in each.

What is the difference between *temperature* and *quantity* of heat? What is the *specific* heat of lead? What are some of the sources of error in your experiment?

V. LATENT HEAT.

A. *Heat Disappears when Solids Liquefy.*

Take the temperature of some water. Wipe the thermometer dry, and take the temperature of some fine salt.

Mix salt in the water and take the resulting temperature.

Take the temperature of some shaved ice. Mix salt with the ice and take the resulting temperature. Does the ice liquefy faster when mixed with salt than otherwise? While it is changing from a solid to a liquid, how is its temperature changing? Why is salt put upon icy pavements? Freezing mixtures.

B. *Heat Disappears when Liquids Vaporize.*

Take the temperature of ether in the bottle. Put a little in a watch crystal or an evaporating dish and note its temperature while it is evaporating. Blow across it (why?) and note the temperature. Put a drop of water into it and see if you can convert it into ice.

Ice machines. The water bath. The double boiler. Animal heat reduced by perspiration. Sprinkling with water for the purpose of cooling. " Muggy days." Humidity. Why is it difficult to cook upon a mountain-top? Freezing produced by reducing the pressure.

C. *Heat Reappears when Vapors Liquefy.*

Take two hundred grams of water in a flask. Note its temperature. Pass steam into it from another flask until its temperature rises—say 50° C. Weigh it to find how much water has been added by condensation of steam. How many units of heat have been added to the flask? How many are due to the liquefying of the steam? How many units of heat for each unit of steam liquefied? Steam heating of buildings. Temperature in the center of a rainstorm.

D. *Heat Reappears when Liquids Solidify.*

Put twenty-seven grams of anhydrous sodium sulphate in fifty grams of water in a flask and heat to 34° C., stirring until all is dissolved. Allow it to cool without depositing crystals. This may be done if it stands very quietly. Now put a thermometer into this liquid and stir. Note the rise in temperature as the sodium sulphate deposits.

Temperature in the center of a rainstorm; tubs of water put in a cellar to keep vegetables from freezing; large bodies of water modify the climate in freezing.

IV. Light.

I. *How Light Spreads from a Center.*—How light decreases in intensity with distance. Measurement of the intensity of light. The visual angle. How we estimate distances. Brilliant objects deceive us as to their size. The sun's light and heat upon slopes.

II. *Shadows*—umbra and penumbra, eclipses.

III. *Light through small apertures*—a pin-hole camera.

IV. *Reflection* from plane mirrors. The apparent location of the image. Reflection from curved mirrors. Enlarged images formed in concave mirrors. Inverted images formed in concave mirrors. Diminished images formed in a convex mirror. A curved image from a straight object. Focusing light by a concave mirror. Mirrors which send out parallel rays. How daylight is diffused. Halos. Brilliant clouds. Phases of the moon. Twilight.

V. *Refraction of Light.*—Index of refraction for water, glass, etc. Some invisible things are made visible by refraction. Enlarged images produced by refraction—lenses. Inverted images produced by refraction—conjugate foci and principal focus. Focal distance of a lens. How pictures are formed at the focus of a lens. To determine the magnifying power of a lens. The spectrum.

V. Magnetism and Electricity.

MAGNETS.

Dip a piece of magnetite into iron filings.   Find whether the filings cling to opposite parts.

Suspend a piece of magnetite.   Note whether one part of it is inclined to point north.

Find a spot on the magnetite where iron filings cling most abundantly, and lay this spot upon one end of a *steel* needle.   Repeat this several times, always in the same direction.   By means of the iron filings find whether the needle has become a magnet.   Try the same experiment using a piece of soft iron wire in place of the needle.

Having magnetized a needle find out whether this needle will magnetize other needles.

Induced Magnetism.

While holding one end of a bar magnet near one end of a short rod of iron, test the power of the latter to lift iron filings.   Remove the magnet.   How is the iron affected?

Bring a horseshoe magnet near to a suspended rod of soft iron so that a pole of the magnet is underneath each end of the iron rod.   Now by means of a bar magnet test the iron rod for polarity.

Fix a small glass tube in a cork, stand it on end, and fill it with coarse iron filings.   Present one pole of a bar magnet to the upper end of the tube and find how long a chain of filings it will lift.   While doing this note the effect of bringing the opposite pole of another bar magnet against the side of the glass tube about two inches from the first magnet.   Try the effect of the like pole of the second magnet.

The Molecular Theory of Magnets.

Magnetize a sewing needle, present it to iron filings. Heat it red hot; now again test its magnetism by the iron filings.

Magnetize a knitting needle; test its magnetism. Drop it upon the floor several times; now test its magnetism.

Magnetize a sewing needle; test it for poles and a neutral point. Break it at the neutral point; test for new poles and neutral points. Break again at neutral points and test for new poles. Place magnets under panes of glass and produce upon the glass figures with iron filings showing the lines of force. Trace such lines of force in the vicinity of magnets, using a small compass.

STATIC ELECTRICITY.

Make a stick of sealing wax, slowly approach and touch a pith ball suspended upon a silk thread. Now do the same after having rubbed the sealing wax with flannel. By sliding the silk thread make this pith ball slowly approach and touch another pith ball suspended in the same manner. Make the stick of sealing wax slowly approach and touch, if it will, each of the pith balls again. Be careful not to touch either of the pith balls with your hand during the exercise.

(Before beginning this exercise touch the pith balls with your hand to discharge any electricity which may have been left in them from the last exercise.) Repeat the foregoing exercise, using a glass tube and silk in place of the stick of sealing wax and flannel.

Discharge with the hand if necessary the two pith balls. Suspend them several inches apart. Electrify one by means of the sealing wax and the other by means of the glass tube. Now make the glass tube approach, *but not touch*, the first pith ball, and the stick of sealing wax approach, *but not touch*, the second pith ball. Also make the pith balls approach each other by sliding the silk threads.

Charge one pith ball positively and the other negatively. Rub the glass tube with a silk cap, and by means of a silk thread make it approach, *but not touch*, each pith ball in turn. Is the silk cap electrified? How? What is the proof?

Rub the stick of sealing wax with a flannel cap, and by means of a silk thread make it approach, but not touch, each pith ball in turn. Is the flannel cap electrified? How? What is the proof?

Conduction.

Suspend a bare copper wire horizontally by silk threads a few inches long at either end. Suspend two pith balls by a few inches of very fine copper wire (No. 36). The upper end of each piece of fine wire is bent into a hook to hang it upon the horizontal wire. Suspend two other pith balls by silk threads from the horizontal wire.

Electrify a stick of sealing wax and touch it to the silk thread at either end of the horizontal wire. Does either pair of pith balls separate?

Slide the stick of sealing wax over to the copper wire. Does either pair of pith balls separate now?

Moisten the silk threads which suspend one pair of pith balls. Try now to conduct electricity to them by touching the copper wire with the electrified stick of sealing wax.

Induction.

Bring an electrified stick of sealing wax underneath one of the pith balls suspended by copper wire. Make the sealing wax approach very near to the pith ball, but do not allow them to touch. While holding the sealing wax near to this pith ball, lift the other pith ball, which is suspended by a copper wire, off the horizontal copper wire and hang it upon the silk thread at one end. A loose piece of silk thread is provided so that this may be done without touching the copper with the fingers. Now test each ball with both glass and sealing wax to determine whether it is positively or negatively electrified.

Hold an electrified stick of sealing wax near without touching a pith ball suspended by a silk thread. Touch the pith ball with your finger while it is near the sealing wax. Test now the pith ball with both glass and sealing wax to determine whether it is positively or negatively electrified.

Can you by means of an electrified glass tube charge a pith ball suspended upon a silk thread either positively or negatively at pleasure?

Electrical Distribution.

Hold a glass tube by the hand placed midway between the two ends; rub one end with silk. By means of a positively electrified pith ball determine whether the other end of the glass is also electrified.

Electrify, by touching it several times with an electrified glass tube, a small tin cup insulated upon a glass beaker. Test both the inner and outer surface of the cup by means of a positively electrified pith ball. Compare also the intensity of the charge in the handle with that upon the outer surface of the cup.

Application of Principles.

Use an electrophorus to charge a Leyden jar so that its outer coat shall be positively charged; negatively charged.

Pass ten or twenty sparks into a Leyden jar from an electrophorus, and then discharge it through your body.

Light the gas at the Bunsen burner by means of a spark from a Leyden jar.

CURRENT ELECTRICITY.

The Battery Cell.

Dip a strip of zinc into dilute sulphuric acid (one volume of acid to twenty volumes of water). Note the bubbles of hydrogen which form upon its surface. Dip a strip of copper into the acid and note that bubbles do not form upon its surface. Allow the ends of the strips which extend above the liquid to touch one another. Note that bubbles of gas now form also upon the strip of copper. Amalgamate the zinc with mercury. Note that few, if any, hydrogen bubbles are now produced upon the surface of the zinc, but when the upper ends of the zinc and copper touch, hydrogen bubbles appear upon the latter. Drop sodium bichromate into the solution until hydrogen bubbles disappear from the copper strip. If the strips are allowed to remain in the acid with the upper ends connected it will be found that the zinc wastes away, while the copper remains undiminished.

Examine a Samson cell and explain how it fulfills the essential requirements of a battery cell.

To Detect the Presence of an Electric Current and its Direction.

Place the cell so that the carbon pole is due north of the zinc pole. Connect them by a short piece of copper wire which will lie in the magnetic meridian. Hold a compass needle over this wire and note in which direction its north-seeking pole is deflected. Hold the compass needle underneath the wire and note in which direction the north-seeking pole is deflected. Turn the cell about so that the carbon pole shall be due south of the zinc pole. Hold the compass needle above the wire and note in which direction the north-seeking pole is deflected. Hold the needle underneath the wire and note in which direction

the north-seeking pole is deflected.   Do not leave the copper wire in circuit longer than is necessary.

By means of the compass needle determine which bind-
ing post at your table is connected with the carbon ele-
ment of the battery.   Label it as the positive pole.

SOME EFFECTS OF ELECTRIC CURRENTS.

A. *Chemical Effects.*

Using carbon terminals for your wires, dip them into
a solution of copper sulphate, not allowing them to touch
one another.   Close the circuit through the push button.
Notice what occurs on each electrode.

Repeat the exercise, and immediately after disconnect-
ing the battery place an electric bell in the circuit in its
stead.

Spread upon a sheet of metal a piece of paper moistened
in a solution of starch and potassium iodide.   Press one
terminal against the metal sheet, and with the other draw
some characters upon the moistened paper.

B. *Heating Effects.*

Send the current through fine iron wire wound around
the bulb of a thermometer, or touching a cake of paraffin,
or with a drop of paraffin on it.

Note the temperature of the solution in a battery cell
before and after using the current.

C. *Magnetic Effects.*

While passing the current through a small helix of wire,
test it for magnetism by bringing it near to a magnetic
needle.

Place the iron core inside the helix and close the circuit.
For evidence that the iron core is magnetized present it
to some bits of iron.

Label one end of your helix *N*, and decide which end of the wire must be connected with the positive pole. Label this wire *X*. Send the electric current through the helix, and by means of the compass needle determine whether your prediction is correct. Record your prediction before the experiment and the result afterward. Place the iron core inside the helix and repeat the experiment.

Electrical Measurements.

In all measurements throw in enough extra resistance to bring the deflection of the needle below 45°.

Measure by means of the micrometer the diameter of each of the specimens of wire furnished you, and determine its number by referring to the table (below). Test your results by the wire gauge.

Diameter of Wire. Unit, One Thousandth of an Inch.

No. 12.... 80	No. 16.... 51	No. 20.... 32	No. 24.... 20	No. 28.... 13
No. 13.... 72	No. 17.... 45	No. 21.... 28	No. 25.... 18	No. 29.... 11
No. 14.... 64	No. 18.... 40	No. 22.... 25	No. 26.... 16	No 30.... 10
No. 15.... 57	No. 19.... 36	No. 23.... 23	No. 27.... 14	

The spools furnished contain copper and German silver wire, Nos. 24 and 30. Find the resistance and length of the wire upon each spool.

Calculate the resistance of your galvanometer. The wire is No. 22. Measure the diameter of the coil, and from that compute the length of the fifteen turns of wire. Test your result by using another galvanometer. Place the needles far enough apart so as not to influence one another.

Find the internal resistance of your battery cell by connecting it with your galvanometer and noting the deflection of the needle. Then introduce resistance until only one half as much current passes (refer to the table of tangents on next page). If you have reduced the current one half you must have doubled the resistance. Take into account the resistance of the galvanometer.

Table of Tangents.

Deg.	Tangents.	Deg.	Tangents.	Deg.	Tangents.	Deg.	Tangents.
1	.017	24	.445	47	1.07	70	2.75
2	.035	25	.466	48	1.11	71	2.90
3	.052	26	.488	49	1.15	72	3.08
4	.070	27	.510	50	1.19	73	3.27
5	.087	28	.532	51	1.23	74	3.49
6	.105	29	.554	52	1.28	75	3.73
7	.123	30	.577	53	1.33	76	4.01
8	.141	31	.601	54	1.38	77	4.33
9	.158	32	.625	55	1.43	78	4.70
10	.176	33	.649	56	1.48	79	5.15
11	.194	34	.675	57	1.54	80	5.67
12	.213	35	.700	58	1.60	81	6.31
13	.231	36	.727	59	1.66	82	7.12
14	.249	37	.754	60	1.73	83	8.14
15	.268	38	.781	61	1.80	84	9.51
16	.287	39	.810	62	1.88	85	11.43
17	.306	40	.839	63	1.96	86	14.30
18	.305	41	.869	64	2.05	87	19.08
19	.344	42	.900	65	2.14	88	28.64
20	.354	43	.933	66	2.25	89	57.29
21	.384	44	.966	67	2.36	90	Infinite.
22	.404	45	1.000	68	2.48		
23	.424	46	1.036	69	2.61		

Make a table showing what strength of current is indicated by the deflection of your needle for each degree from 85° to 36°, when fifteen turns of wire are used upon your galvanometer; also when five turns are used.

Make a table showing the electro-motive force of your cell in each of the following cases: Use fifteen turns of wire. Close the circuit through the galvanometer alone; add half an ohm of resistance; add one ohm; add two, three, four, etc., to ten ohms. Does the E. M. F. vary with the resistance?

Test your cell with the voltmeter instead of the galvanometer, adding from zero to ten ohms of resistance.

Choose two cells which are as nearly alike as possible in E. M. F. Connect them in series, and first calculate as near as you may the amount of current which they will send through five turns of wire on the galvanometer and two ohms of added resistance; then close the circuit, and find how the result compares with your estimate.

Connect the cells in parallel and repeat the foregoing exercise.

Use the ammeter in place of the galvanometer, and find out whether these two cells send more current through it when connected in series or in parallel: (*a*) without added resistance; (*b*) with added resistance of .5 ohms.

Find about what current is required to operate the pieces of apparatus furnished you. What resistance does each offer? What arrangement of battery is desirable?

Connect two motors with a rubber band. Operate one with a battery and connect the other with the galvanometer to see whether it is producing a current as a dynamo. If so, how much? What is its electro-motive force? How much current is used in the first motor?

VI. Sound.

i. *How Sound is Produced.*

All sound-producing instruments are in a state of vibration, and cause vibration in the air.

While a tuning fork is producing a sound touch the prongs to the surface of water; touch them to the tip of your tongue; touch them to the edge of an empty tumbler.

Bring a pith ball aginst a piano string while it is sounding; against the rim of a bell while it is sounding.

Cause an organ pipe to sound near the end of a long glass tube containing cork dust.

Strike a bell while holding it a few inches above a diaphragm upon which cork dust has been sprinkled.

ii. *How Sound is Transmitted.*

In order that sound may be heard, air or some medium must extend all the way from the sounding body to the ear.

Ring an electric buzzer suspended in a bottle; exhaust

the air and ring again; allow the air to enter and ring again; pour in some ether and after it has evaporated ring again; fill the bottle with hydrogen and ring again.

Submerge a tumbler in a pail of water, and while holding it suspended beneath the surface strike it with a piece of metal so as to cause it to ring. The sound is transmitted through the water to the air, and through the air to the ear.

Fasten the opposite ends of a long wire into the bottoms of cigar boxes or large pasteboard boxes, and hold one box against the ear while some one taps upon the box at the farther end of the wire. Hold the wire suspended freely throughout its entire length. The acoustic telephone.

III. *Reflection of Sound.—Velocity of Sound.*

If possible, take an excursion to some place where a good echo may be obtained. Try to obtain from it also data for calculating the velocity of sound.

IV. *Intensity and Loudness of Sound.*

Amplitude of vibration; we strike the tuning fork or piano string harder to make a louder sound; density of medium; refer to the electric buzzer in bottle; distance; area of vibrating body; cf. violin and violoncello; resonators.

V. *Interference.—Sympathetic Vibrations.*

Why may the voice be heard better in a moderate-sized room than out of doors? Why may it be more difficult to hear a voice in a large room than out of doors? Try the usual experiments with tuning forks to illustrate interference. Open a piano so as to expose the sounding board, lift the damper from a string, and sing the tone which that string produces. Find out whether the string is vibrating in sympathy.

vi. *Pitch.—The Scale.*

Draw a visiting card across the teeth of a saw to illustrate that the pitch of a sound rises as the vibrations become more rapid. Try the usual experiment with cardboard siren. What effect has shortening the prongs of a tuning fork upon its pitch? How does weighting the prongs affect the pitch?

vii. *Vibration of Strings.*

With a simple sonometer verify the laws of length, of thickness, and of tension.

viii. *Vibrations of Air Column in Organ Pipes.*

Have a series of small organ pipes and determine how the pitch is related to the length of the pipes; also the relation of open and closed pipes.

Syllabus of Topics for Lecture and Recitation Work in Physics for the Second High-school Class.

I. *Matter, Force, Mechanics of Solids.*

Familiar illustrations that will develop ideas of the molecular constitution of matter and of molecular force, including such topics as the solid, liquid, and gaseous states, and the relation of heat to these; cohesion; crystallization; capillarity; diffusion and osmose.

Simple and compound substances; physical and chemical changes; indestructibility of matter; the correlation of physics and chemistry.

Familiar illustrations and applications of the principles of momentum, gravitation, and the three laws of motion; the composition of parallel forces and the application of the principle to levers, pulleys, muscular movements, the wheel and axle, balances, devices for illustrating center of gravity and the stability of bodies; uniformly accelerated and retarded motion; the path of projectiles; the pendulum and its uses; the recoil of a gun; the motion of the rotary lawn sprinkler.

II. *Dynamics of Fluids.*

Pressure in liquids and gases; the barometer and its applications; transmission of the pressure exerted on fluids.

Application of principles to the siphon, pumps, springs, and artesian wells; methods of supplying cities with water; the hydrostatic press; pneumatic tubes.

Discussion of buoyancy and specific gravity.

III. *Heat.*

Sources of heat such as friction and chemical action; effects of heat on solids, liquids, and gases in producing expansion, liquefaction, and vaporization; measurement of heat with reference to temperature and quantity; radiation and diffusion of heat in solids, liquids, and gases by conduction and convection.

Discussion of principles as illustrated in such topics as ventilation; heating of houses by steam, hot water, and hot air; winds; ocean currents; freezing mixtures; animal heat; weather phenomena.

IV. *Light.*

Energy of light; methods of comparing the illuminating powers of two sources of light; explanation of images and shadows; umbra and penumbra; reflection of light; refraction of light; mirrors, prisms, and lenses; the spectrum; color; correlation of heat, light, and chemical rays.

Application of principles to simple and compound microscopes, the telescope, the stereopticon, the eye, the rainbow, eclipses, phases of the moon, etc.

V. *Magnetism and Electricity.*

Magnets. Polarization; attraction and repulsion; magnetic induction; the magnetic needle; the dipping needle; the earth as a magnet.

Statical charges. Attraction and repulsion; two kinds of electrification; conductivity; insulation; induction; effects of points; lightning.

Current electricity. The battery cell; effects of the current on a magnetic needle; chemical action produced by the current; heat produced by the current; electro-magnets; electric measurements; induced currents.

Application of principles to the telegraph, the electric bell, the telephone, the storage battery, the arc light, the incandescent light, the dynamo, the electric motor, and electroplating.

VI. *Sound.*

Production of sound; its transmission in solids, liquids, and gases; velocity of sound; reflection of sound; echoes; re-enforcement of sound; interference; nodes and antinodes; forced and sympathetic vibrations; intensity, pitch, and quality of sound.

Application of principles to the vocal organs, the ear, and acoustic instruments.

SUGGESTIVE EXAMINATION QUESTIONS.

1. A cubic foot of ice, weighing fifty-six and a quarter pounds, is floating in water and one tenth of its volume rises above the surface of the water. Show how you may compute from this the weight of a cubic foot of water.

2. Granite is about two and a half times as heavy as water; how much would you be obliged to lift while holding a cubic foot of it submerged in water?

3. Suppose a body, weighing ten pounds, moves under the influence of a constant force twenty feet the first second, sixty feet the next second, and one hundred feet the third second, how far will it move the fourth sec-

ond? How does its momentum at the end of the fourth second compare with that at the end of the second second? Explain your method of calculation.

4. Explain why a stream of water falling freely tends to separate into drops.

5. In a certain wheel and axle the diameter of the wheel is four feet and that of the axle is one foot. It is found that a weight of one hundred and sixty pounds attached to the rope which encircles the wheel is necessary to lift a weight of six hundred pounds attached to the rope which encircles the axle. How much work must a man do to lift the six hundred pounds a distance of two feet with this machine? How much work would he need to do to lift it without the machine?

6. Explain how it is that snow protects vegetation in winter. Will wrapping a thermometer in woolen make it warmer? Why then do woolen blankets keep the body warm?

7. How do large bodies of water modify the climate about them? Explain.

8. Why is it warmer in the center of a storm than on its outskirts?

9. Why does the skin feel cold when moistened with alcohol? Why does it feel cooler after being moistened with warm water than with cold water?

10. Why does a mixture of ice and salt freeze water, and why does the ice liquefy at the same time?

11. Draw a diagram and explain how a curved mirror may give an inverted image.

12. Draw a diagram and explain how ripples upon the surface of water prevent one from seeing objects submerged in it.

13. Draw a diagram and explain how a tumbler of water may magnify things seen through it.

14. With a permanent magnet explain how you would magnetize a sewing needle so that its point would repel the north end of a compass needle.

15. How would you accomplish the same thing, using instead of the permanent magnet insulated wire connected with a battery cell?

16. How may you, with a stick of sealing wax, electrify one pith ball positively and the other negatively?

17. Draw a diagram of an electric bell and show how an electric current operates it.

18. Show how an electric bell may be used to give a shock.

19. Distinguish between pitch and loudness of sound.

20. If two vibrating strings are alike in all respects, except that one is twice as long as the other, how do their tones differ in pitch? Answer the question in case they are the same length, but one has twice the diameter of the other.